BUILDING BLOCKS OF CHEMISTRY

CHEMISTRY OF LIVING THINGS

Written by William D. Adams

Illustrated by Maxine Lee-Mackie

WORLD BOOK

a Scott Fetzer company
Chicago

World Book, Inc.
180 North LaSalle Street
Suite 900
Chicago, Illinois 60601
USA

For information about other World Book publications, visit our website at **www.worldbook.com** or call **1-800-WORLDBK (967-5325)**.

For information about sales to schools and libraries, call 1-800-975-3250 (United States), or 1-800-837-5365 (Canada).

© 2023 World Book, Inc. All rights reserved. This volume may not be reproduced in whole or in part in any form without prior written permission from the publisher.

WORLD BOOK and the GLOBE DEVICE are registered trademarks or trademarks of World Book, Inc.

Library of Congress Cataloging-in-Publication Data for this volume has been applied for.

Building Blocks of Chemistry
ISBN: 978-0-7166-4371-5 (set, hc.)

Chemistry of Living Things
ISBN: 978-0-7166-4379-1 (hc.)

Also available as:
ISBN: 978-0-7166-4389-0 (e-book)

Printed in India by Thomson Press (India) Limited, Uttar Pradesh, India
1st printing June 2022

WORLD BOOK STAFF

Executive Committee
President: Geoff Broderick
Vice President, Editorial: Tom Evans
Vice President, Finance: Donald D. Keller
Vice President, Marketing: Jean Lin
Vice President, International: Eddy Kisman
Vice President, Technology: Jason Dole
Director, Human Resources: Bev Ecker

Editorial
Manager, New Content: Jeff De La Rosa
Associate Manager, New Product: Nicholas Kilzer
Proofreader: Nathalie Strassheim

Graphics and Design
Sr. Visual Communications Designer: Melanie Bender
Sr. Web Designer/Digital Media Developer: Matt Carrington

Acknowledgments:
Writer: William D. Adams
Illustrator: Maxine Lee-Mackie/ The Bright Agency
Series Advisor: Marjorie Frank

TABLE OF CONTENTS

Introduction 4

Elements of Living Things 8

Functional Groups in Biomolecules 14

Nucleic Acids 16

Carbohydrates 22

Lipids 24

Amino Acids:
The Building Blocks of Proteins 27

Enzymes 31

ATP and Cellular Respiration 33

Photosynthesis 36

Conclusion 38

Words to Know 40

There is a glossary on page 40. Terms defined in the glossary are in type **that looks like this** on their first appearance.

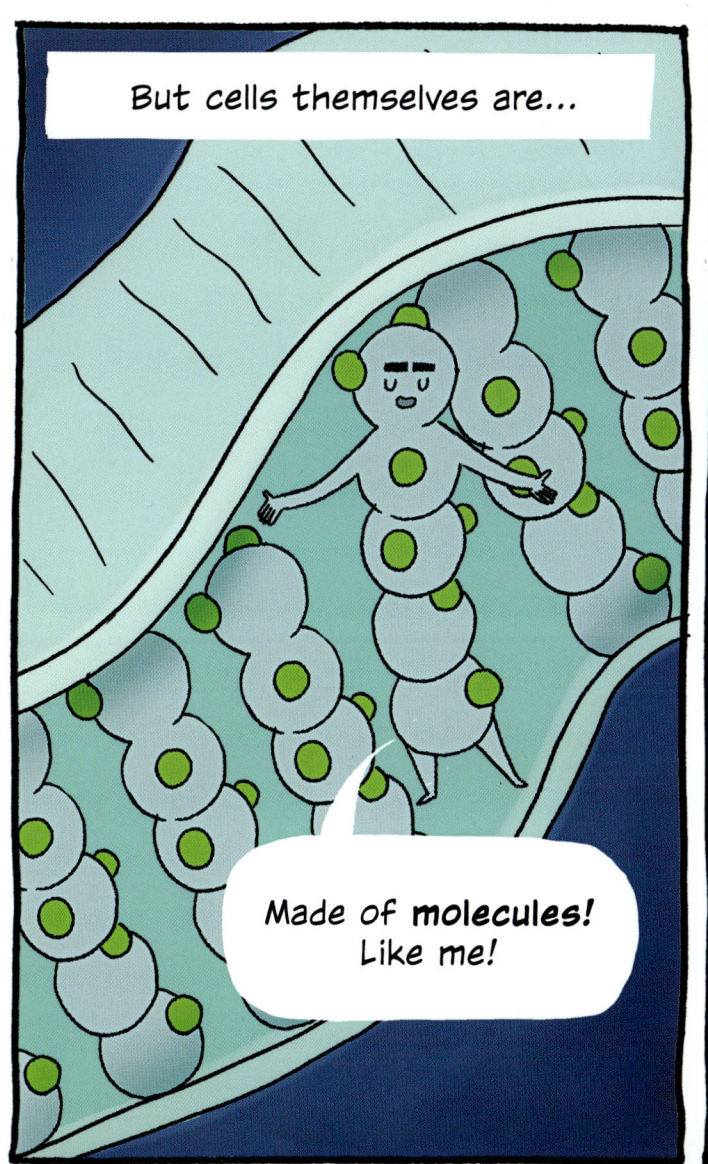

Living things are made of and produce biological molecules or **biomolecules**.

The chemistry inside an organism (living thing) makes it what it is! And chemistry keeps the organism alive and functioning.

ELEMENTS OF LIVING THINGS

There are dozens of types of atoms.

Chemists call each different type of atom a **chemical element**.

A chemical element is a substance that contains only one kind of atom and cannot be broken down into simpler substances.

"PERIODIC TABLE"

Living things only use some of the elements.

But rocks aren't alive—and neither are air and water.

Living things get the elements they need from the food, water, and air they take in.

Why are carbon, nitrogen, oxygen, and hydrogen so important to life anyway?

Carbon, nitrogen, and oxygen can form multiple chemical bonds. A bond is a strong force that holds atoms together.

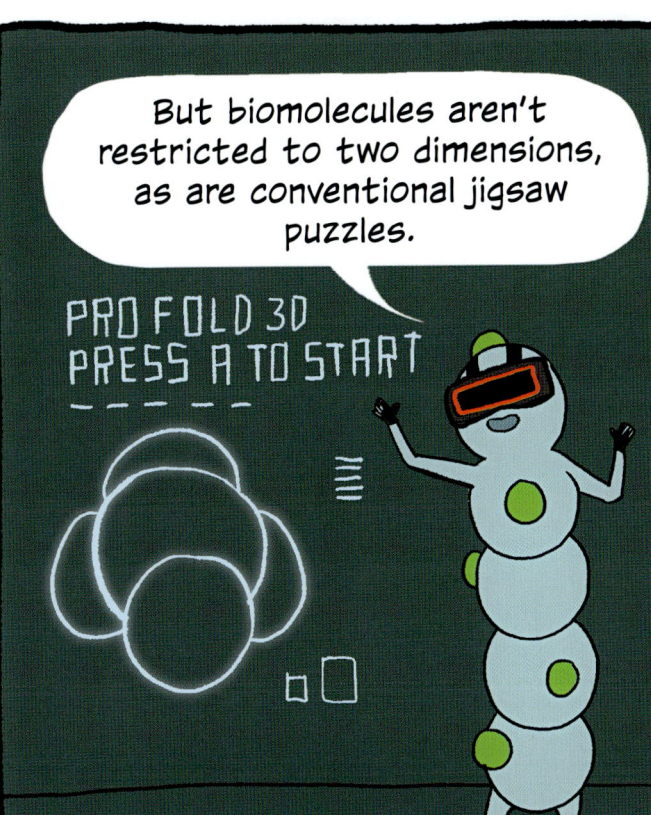

Such pieces are called functional groups. A functional group is a group of atoms that gives rise to the characteristic properties of certain compounds.

Functional groups are kind of like the standardized parts used to manufacture bicycles and other vehicles.

TEST AREA!

Hmm...

Wherever the functional group goes, it will always have the same job and chemical behavior.

Functional groups and assorted atoms bond together to make four main types of biomolecules: nucleic acids, carbohydrates, lipids, and proteins. Let's check them out!

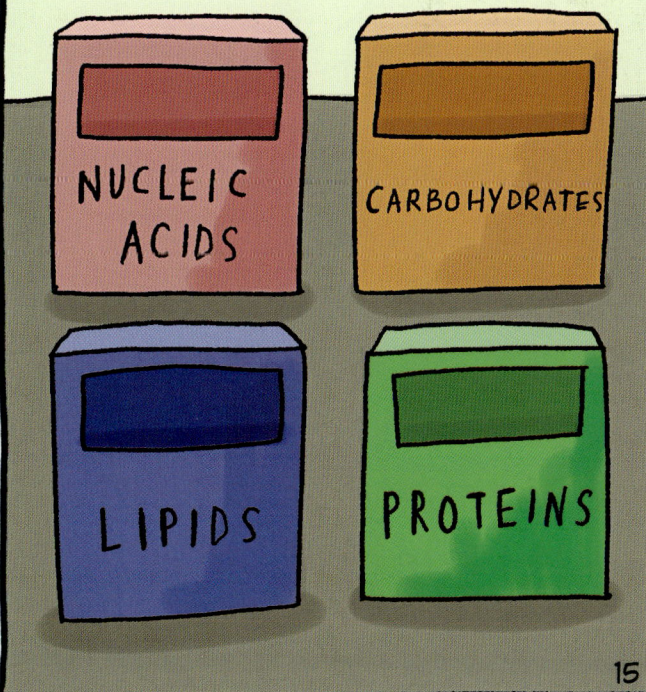

NUCLEIC ACIDS

CARBOHYDRATES

LIPIDS

PROTEINS

15

Nucleic acids carry instructions that tell cells when and how to make the proteins they need to survive and grow.

Nucleic acids are made of smaller chemical units called **nucleotides**. Each nucleotide in turn consists of three functional groups: a phosphate, a sugar, and one of a special group of molecules called *base compounds*, or just **bases** for short.

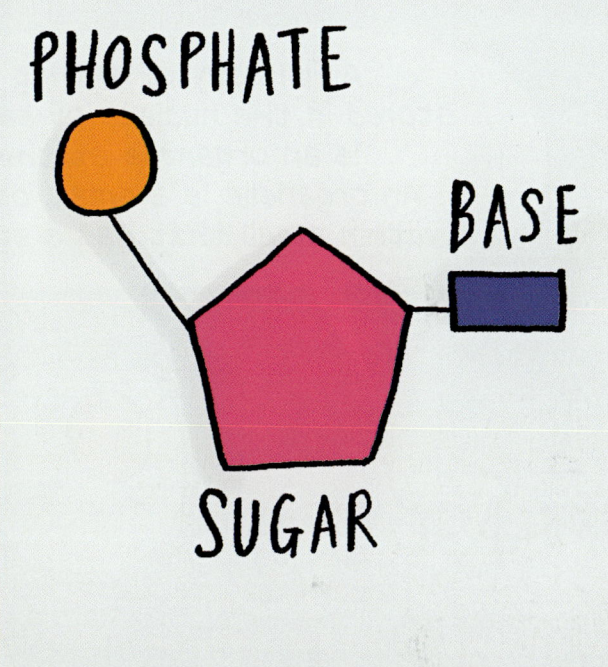

Each base stores one single piece of information. But you can't write a set of instructions with one letter! So, the phosphate and sugar groups of different nucleic acids can link together to form long chains.

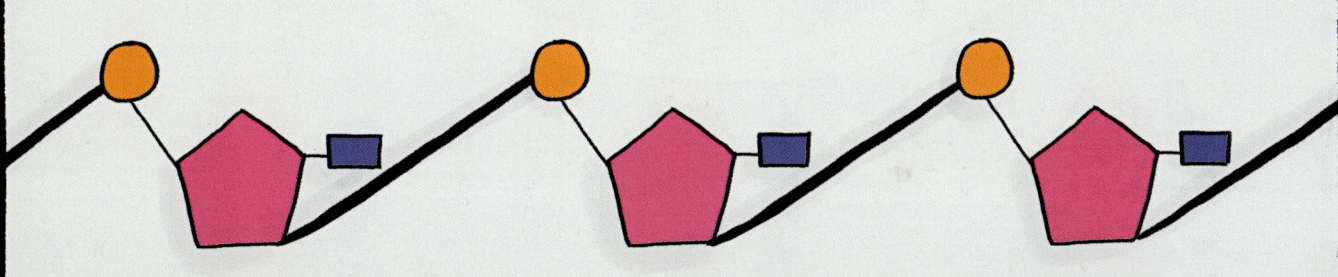

17

A strand of DNA holds thousands of sections called **genes**.

Genes are the units of code that determine every *trait* (characteristic) passed on to living things by their parents.

Genes control everything from the color of a flower to the size of an animal's nose.

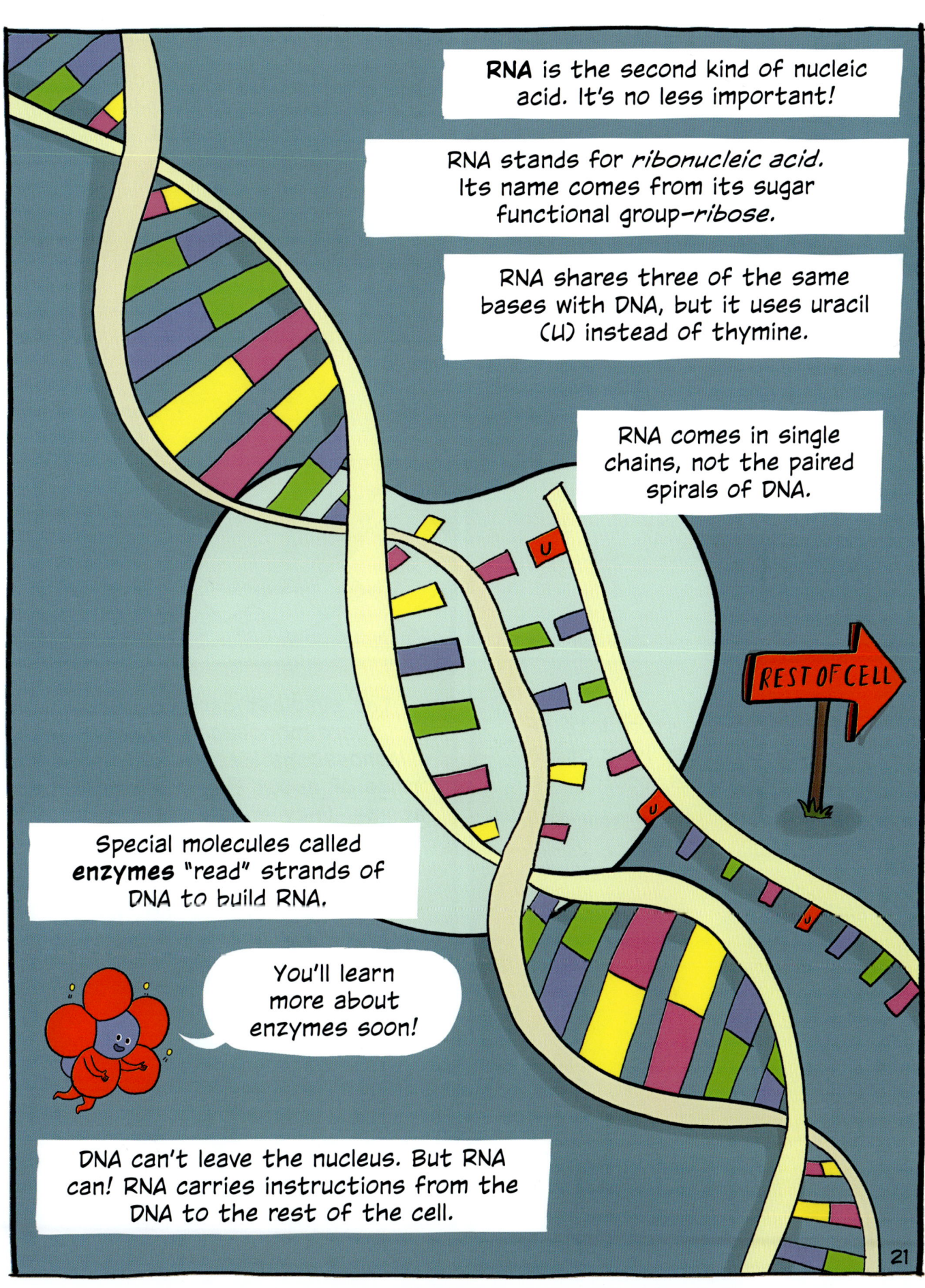

Monosaccharides can easily bond with each other. Two monosaccharides bonded together make a disaccharide.

Disaccharides make up table sugar and the sugar found in dairy products. They are also found in fruits and vegetables.

Monosaccharides can also link together in long chains. Such biomolecules are called polysaccharides. A polysaccharide called cellulose gives strength and structure to roots, stems, and vines.

Polysaccharides can be good to eat, too. They are found in breads, rice, and other starchy foods.

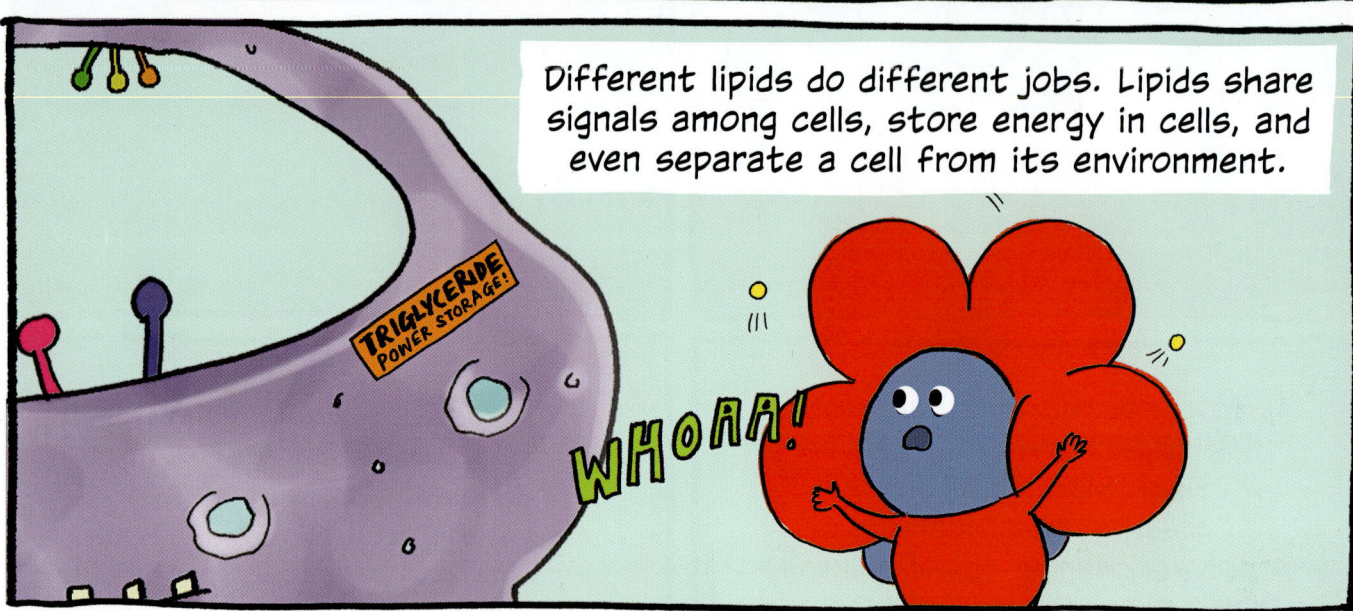

Each phospholipid has a phosphate functional group at one end and two **fatty acid chains** on the other end. The phosphate "head" is *hydrophilic*—that means it's attracted to water. But the fatty acid chains are *hydrophobic*—they're repelled by water.

Think what would happen if some of these phospholipid molecules were surrounded by water molecules.

They automatically arrange themselves so that their heads point out toward water and water-based solutions and their chains point in. Together, the heads attract the water and keep it from getting to the chains.

But now imagine there are a *bunch* of phospholipid molecules in the water.

There are too many molecules to form a little bubble, so instead the phospholipids arrange in two layers facing opposite directions—a **lipid bilayer**. The tails still cluster together away from the water, but many more molecules can gather, forming a broad "sheet."

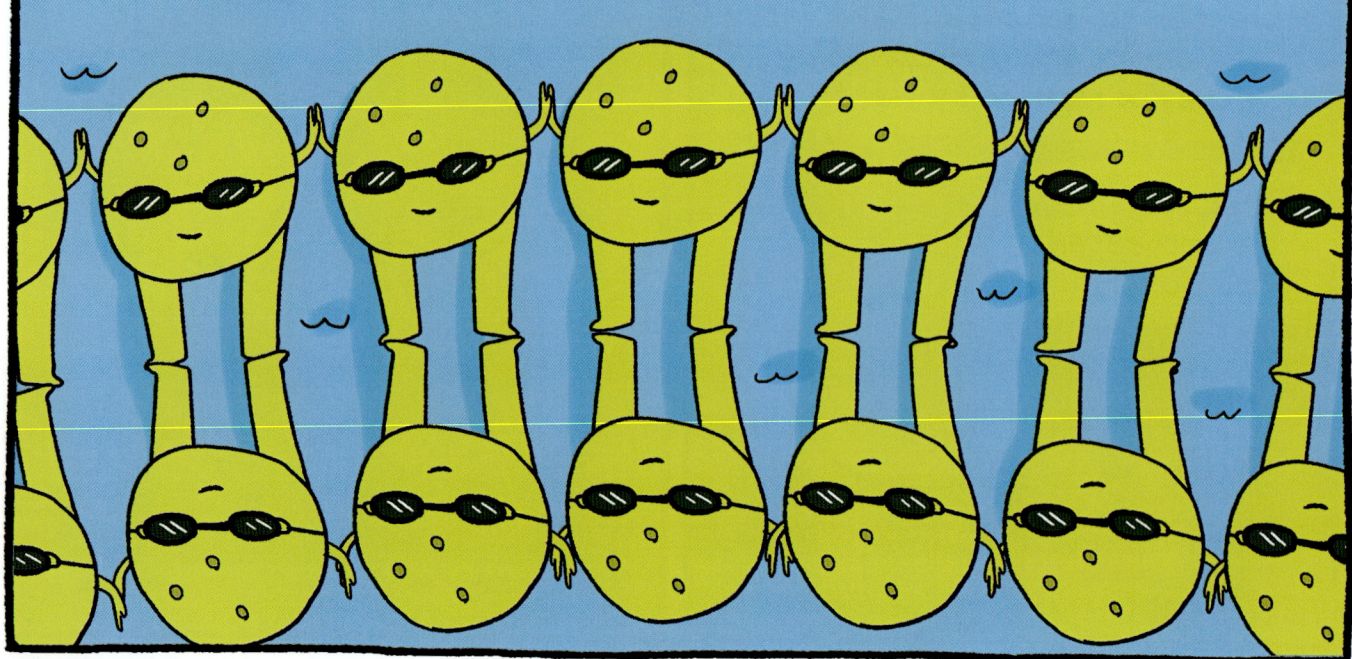

Just as a soap bubble encloses a little bit of air, a lipid bilayer bubble can enclose a little bit of liquid.

Lipid bilayers form cellular membranes, acting as the "skin" of a living cell. They keep cytoplasm inside the cell and water and other fluids out.

The cell assembles amino acids into long chains called proteins. The amino group of one amino acid links up with the carboxyl group of another.

Depending on the amino acids that make up a protein chain, the chain can coil and wind in different ways.

The chain can also attach to other protein chains.

ENZYMES

Living things rely on a variety of chemical reactions. Many of these would happen much too slowly—or wouldn't happen at all—without help.

But chemical reactions do have help! Some proteins belong to a family of helper biomolecules called enzymes.

Enzymes are molecules that speed up chemical reactions in all living things.

Each cell contains a bewildering number of enzymes that allow it to perform its function in the body, repair itself, and reproduce.

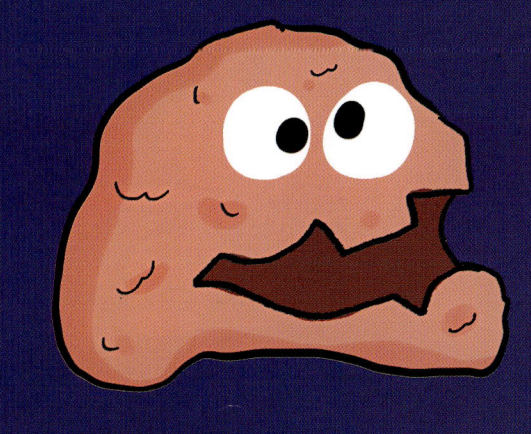

ATP AND CELLULAR RESPIRATION

Where does the energy for such a reaction—or any cellular process—come from?

In cells, energy is stored in a special molecule called adenosine triphosphate—ATP. It gets its name from the three phosphate groups that are linked with its other functional groups.

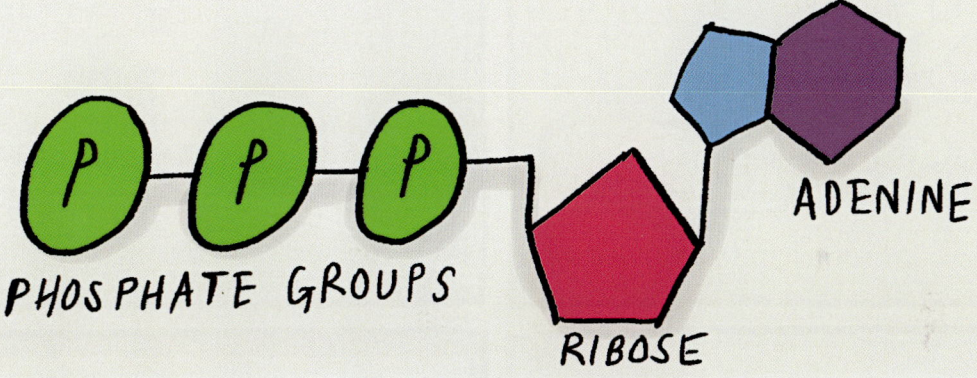

Each ATP is like a tiny battery! When that farthest phosphate group splits off from the molecule, it releases lots of energy.

Mitochondria combine molecules from food and oxygen from the air to produce energy. This process is called **cellular respiration**.

Water and carbon dioxide are also produced in this chemical reaction.

The energy produced in cellular respiration converts the used-up adenosine diphosphate (ADP) and phosphate groups back into ATP.

The ATP flows back out into the cytoplasm, ready to be used for other cellular functions.

PHOTOSYNTHESIS

"Cellular respiration is great, but what if you're a plant and don't really eat?"

uh-uh

Plants have a little trick up their leaves called **photosynthesis**.

Photosynthesis

Through photosynthesis, plants convert carbon dioxide from the air and water from the ground into sugars using sunlight!

Light is a form of energy. Organelles called **chloroplasts** use this energy to combine carbon dioxide and water molecules, which normally aren't very reactive.

"Sorry, I must be leaf-ing!"

A plant uses some of these sugars to build its tissues.

It breaks down the rest to make energy via cellular respiration.

SCORE!

That means plants still need oxygen...

...but they produce more than they need through photosynthesis. They release the extra oxygen into the air!

WORDS TO KNOW

amino acid any one of a group of compounds of nitrogen, hydrogen, carbon, and oxygen that combine in various ways to form the proteins that make up living matter.

atom one of the basic units of matter.

base any of the four chemical compounds—adenine, cytosine, guanine, and thymine—that are present in nucleic acid and combine in various ways to form DNA.

biomolecule any one of certain molecules produced by and found in living things.

cellular respiration a series of chemical reactions that occur in the presence of oxygen. These reactions release energy from food and make it available so that the cells can use it.

chemical element a substance made of only one kind of atom. There are 118 chemical elements.

chloroplast a tiny structure found inside the cells of plants where photosynthesis occurs.

DNA deoxyribonucleic acid, a chainlike molecule found in every living cell. It directs the formation, growth, and reproduction of cells and organisms.

enzyme a molecule that speeds up chemical reactions in all living things.

fatty acid chain a group of certain molecules joined together that makes up the building blocks of fats.

gene a section of DNA that determines which characteristics living things inherit from their parents.

glycolysis a process by which a carbohydrate is broken down to produce usable energy for a living thing.

lipid bilayer a thin membrane composed of two layers of lipid (fat) molecules.

molecule two or more atoms chemically bonded together.

mitochondria parts of a cell that convert chemical energy from food into a form of energy the cell can use.

nucleotide the chemical building block that makes nucleic acids, including DNA.

phospholipid any one of a group of phosphorus-containing lipids (fats). These molecules make up the cell membranes and other cell parts in living things.

photosynthesis the chemical process by which plants make their own food using sunlight, water, and carbon dioxide gas.

RNA ribonucleic acid, a complex molecule that plays a major role in all living cells.